To Becca
Happy Tracking!!!

Library of Congress Control Number: 2015907665
ISBN: 978-1-57833-620-3 (soft cover)
ISBN: 978-1-57833-619-7 (hard cover)

Printed by Everbest Printing Co., Ltd., in Guangzhou, China,
through Alaska Print Brokers, Anchorage, Alaska.

Published by:
Studio 2138
2138 Buffalo Drive
Grand Junction, CO 81507
www.studio2138.com

Dedication

For Susan

My Muse........

With whom all things are possible

Contents

Preface

Imagine a creature walking along a beach, and it wasn't yesterday or last week or a month ago...
It was 200 million years ago.

Then imagine finding remnants of that beach, now turned to stone, eroding out of a cliff. When you examine it, you see the tracks of that beach-walker from so long ago.

Then you realize that you may be the first person ever to see these traces and recognize them for what they really are.

- Andre Delgalvis

Some people say that Bullfrog, Utah got its name from rock a formation that can be seen from State Route 276, the only road leading in and out of the place. I hesitate to call Bullfrog a town, but I guess it could be likened to a company town of years past. It has a post office, laundromat, gas station, a restaurant, its own power supply, a hotel, and worker's living quarters. Until recently, it even had an honorary mayor, John Fraser, although a lot of folks down there knew him by his call sign – Capt. Rimrock -.

I had traveled that road to Bullfrog countless times over the years, and until last year never could find that rock formation. A lot of people I talked to said they knew about it, even described what position the frog was in. Trouble was, when I asked for an exact location a lot of different places were mentioned, none of which provided a convincing image of that frog. Urban legend came to mind...........

Then last year I met Terry Bell from Ticaboo, a part-time natural history interpreter for the National Park Service and dinosaur tracker, who was able to finally solve the mystery. Now every trip to Bullfrog is met with a sighting of that no longer elusive creature.

Bullfrog's main reason for existence is the fact that it is only one of four marinas to service boat traffic on Lake Powell. It is also located near the center point for locating trace fossils including dinosaur tracks found in a formation known as the Glen Canyon Group.

The area around Glen Canyon was one of the last places in the United States to be formally surveyed and mapped. Due in part to its remoteness, challenging terrain, and harsh environment, few permanent population centers became established there.

Before the completion of Glen Canyon dam, much of the travel in the area was done through several of only a handful of routes that allowed passage across the Colorado River. Other than river runners, few people traversed Glen Canyon from end to end. Those who did, for the most part, limited their exploration to the area immediately surrounding the river's shore and side canyons. Steep cliffs on either side of the river's channel limited further exploration and investigation.

After completion of the Glen Canyon dam and the filling of Lake Powell in the 1980's, the water level in places was over 550 feet above the level of the original river channel. At full pool, the lake is a little over 180 miles long but has almost 2,000 miles of shoreline due to the many side canyons it filled. This provided new access to areas previously inaccessible due to the steep cliffs.

During the 1990s, the National Park Service commissioned a study of the paleontology assets of the Glen Canyon Recreation Area. During this time, the lake was at or near its full pool level. That study documented 36 sites bearing the tracks of creatures that roamed the area before and during the of the age of dinosaurs.

Beginning in 2000, a prolonged drought in the West caused lake levels to fall, finally reaching a record low in the spring of 2005 at almost 145 feet below full pool.

Discovery

I am a professional photographer and have been photographing at Lake Powell since the early 1990's. In July of 2003 my wife and I were at the lake testing out a pair of inflatable kayaks, and exploring a small cove. The lake level had been dropping for the last three years and was at the lowest level that I'd ever seen it. While paddling close to the shoreline, I passed a large rock that was partially submerged and for some reason stopped and turned around to look at it.

It was a little past noon and the sun was high in the sky. The rock had two large impressions on it that seemed rather curious. My wife Susan paddled up next to me and we both looked at the slab for a while. Finally, I said "I think those are dinosaur tracks". Susan was a little skeptical, and truth be told so was I. We paddled back to the houseboat, but I kept thinking about that slab for the rest of the afternoon. As evening approached, I decided to go back and photograph it. Now that the light was at a lower angle, shadows outlined the markings making them easier to see.

We both looked at the photographs over and over that evening and finally concluded that yes, these did indeed seem to be dinosaur tracks!

Journal entry
Ship's log of the *Daina Marie*
October 12, 2003:

"On my way back from fishing up-lake, I noticed an unusual big red rock on the shore of main channel. I had found dinosaur tracks in July and have since been on the lookout for more.

As I headed toward the rock, I passed another big boulder that was standing up on end and it had a split down the middle. I looked inside the split and found two tracks that appeared to be of museum quality.

A quick inspection of the rock that I had originally spotted didn't seem to reveal any tracks".

"About a week and a half later, I returned to photograph and explore the area more. A dinosaur trackway and three more tracks on the original split boulder were discovered".

"A second closer inspection of the red rock I originally spotted from the boat, turned out to contain a number of small chicken or turkey sized tracks".

After that episode I began to realize that the unprecedented lower lake levels were revealing rock strata that possibly had never been explored before.

From that time on, my focus at the lake turned from photographing and fishing to actively searching for dinosaur tracks. Over the next year and a half, the lake kept dropping more and more, and I kept finding more and more dinosaur tracks

In the fall of 2004 I did a presentation for a photography club in Grand Junction, Colorado that included some images of the tracks from Lake Powell. Afterwards, one of the participants, asked me if knew George Callison, a retired paleontologist and artist that lived in the area.

I did not know George but was eager to get his contact information to share some of what I had found. When we got together I showed him a number of the photographs I'd taken at the track sites and asked his opinion as to what the possible track makers could be.

George told me that the study of dinosaur tracks was a kind of sub-specialty in paleontology and was not something actively pursued by a majority of paleontologists. However, he had several friends that did specialize in tracks and was kind enough to give me contact information.

Lake Powell is a reservoir impounding the Colorado River. After the Glen Canyon dam was completed, it took almost 20 years for the lake to fill because water had to be constantly released in order for the Colorado River to flow below the dam. The majority of water in the Colorado River comes from snow-melt in the Rocky Mountains.

Winter snow levels in the mountains had been well below average for the past four years. This trend changed dramatically during the winter of 2004 - 2005. That February, projections indicated that once the season's snow melted the lake could rise as much as 50 feet.

If this prediction held true it would mean that a majority of the newly discovered tracks would again be submerged and possibly not be seen again during our lifetime.

At that point I emailed one of the contacts that George Callison had provided. A short time later I received a response from Dr. Martin Lockley, a professor of geology and paleontology, and dinosaur track expert at the University of Colorado at Denver.

As it turned out, Martin was the paleontologist who did the original track study for the Park Service at Lake Powell during the 90's. After a lengthy correspondence we arranged to meet at Bullfrog in March before the runoff would begin, and when the lake would be at its lowest level for the year.

During the five days we were together, there was not enough time to show him all of the tracks sites that I'd found.

On our last day together, I asked Martin, what if any significance there was in the tracks that we had thus far observed. He drew up a list of the following:

1.
Rare, very fragile baby dinosaur tracks found in the Wingate formation are the best seen in this formation in all of Utah.

2.

"The longest trackway in the Western United States, possibly the world, of a track type called Anomoepus".

- Dr. Martin Lockley in background -

3.

"The only front foot print of the track type Anomoepus ever found in the Western United States."

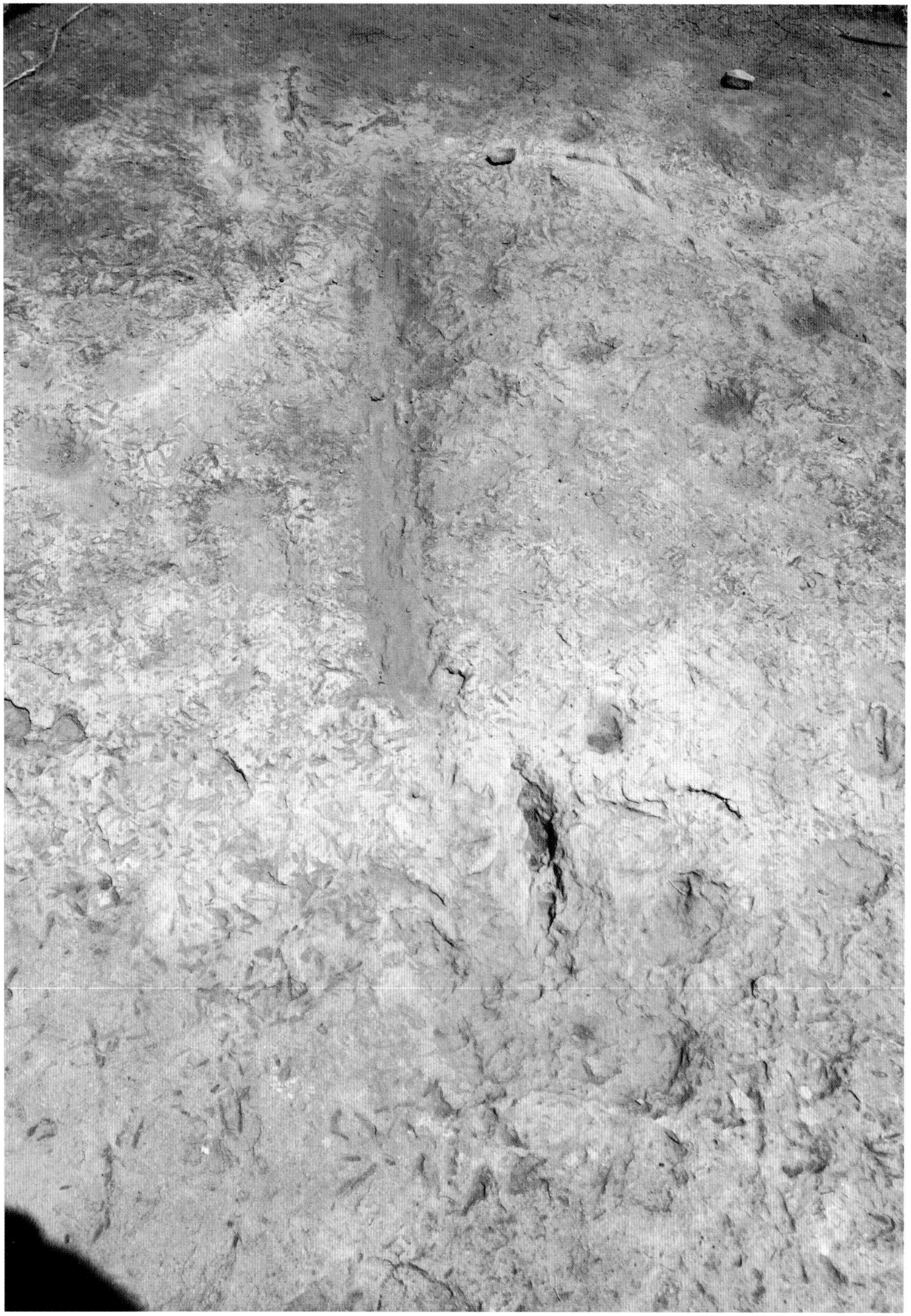

4.

"Dinosaur tail drag marks".

5.

"Scrape marks signifying swimming. Very clear and unusual".

6.
"Tracks of a squatting dinosaur,
the first ever found in the Kayenta formation and the second ever found in Western United States".

7.

"Some of the best preserved tracks of Batrachopus in the West".

(small crocodilian tracks)

8.

"A fantastic abundance of well preserved tracks".

The day after Martin left, I found a new site that contained some of the best, most well preserved tracks and trackways of any I had previously found at the lake.

Time was short, historically the beginning of the snow-melt runoff would reach Lake Powell some time during early to mid-April and from that point on the lake would continue to rise until sometime in late June or July. During this time the lake could rise as much is 1 foot per day.

As a result, Martin and I decided to attempt a collaborative effort to document and/or rescue the dinosaur tracks before they were again submerged. We approached National Park Service in Page, AZ. with a proposal which summarized the scope and significance of the new finds and a proposed recovery scenario. This recovery effort could then possibly be the foundation to establish a permanent display and educational site for the tracks, ideally in to be located in Bullfrog, Utah. While enthusiastic, the Park service apparently lacked the necessary funds in their budget to underwrite the project.

We proposed that it might be possible to raise funds by seeking outside corporate and other community minded sponsors for the project if they were made aware of the urgency of the situation. As a result we contacted the news media with information about the new discoveries.

KSL television from Salt Lake City sent a crew to the Lake, filmed the work we were doing and aired several news stories. A front-page article ran in the Sunday edition of Denver Post as well as in other newspapers in both Colorado and Utah.

LUND

As a result we received numerous offers of help from individuals, including the use of a barge to lift and transport the larger slabs to Bullfrog. Additionally, Martin received a small amount of funding from several outside sources. Aramark Corporation also provided use of a houseboat for a part of the formal documentation stage of the project.

Martin recruited a number of volunteers from his dinosaur trackers research group to come to the lake to help with the project. After we identified the most significant sites, documentation was undertaken. Martin had a permit from the National Park Service to collect and catalog specimens. The smaller ones were initially transported to his dinosaur track museum on the CU Denver campus in Colorado. We were planning on bringing out the larger blocks utilizing the barge, and unloading them for storage and eventual display at Bullfrog.

Unfortunately, on the eve of the barge collection phase of the project, we were notified by the Park Service that no slabs could be stored in Bullfrog for more than 30 days. Because of the size of the slabs it would be impractical to transport them to the Denver museum and because of the rising water time was running out for their rescue. The building that we had hoped to use in Bullfrog for the museum had been turned into a Park Service warehouse so no facilities existed for the tracks to be permanently exhibited. Thus, these blocks were destined to again be covered by the waters of the lake.

TWIN CITIES

May 19, 2005

June 6, 2005

June 29, 2005

By July, the Lake had risen 53 feet. The vast majority the tracks that I had discovered were again underwater. For me, the search for new tracks at the Lake continued as before. The big difference now was that the "easy tracks" near the waterline were for the most part documented. Now the search would have to continue in higher, more difficult terrain.

Over the next few years, many more track sites were located. Arizona Highways Magazine became interested in the finds and published an article in their February 2006 issue. The next year a PBS affiliate from Flagstaff produced a segment about the tracks in a program they aired entitled "Under Arizona".

Since the low water level of the spring of 2005, the Lake level has risen and fallen depending upon the winter snow-pack. It has yet to fall to the point where all of the early tracks sites have been again uncovered.

Most of the track images illustrated in this book have never before been published or seen by the general public.

So how is it that there came to be dinosaur tracks at Lake Powell in the first place?

A number of factors have to be present in order to create and preserve a track site. First you need a track maker to be in existence at the time that a sedimentary deposit is exposed on the surface. A track site recently found in Poland is believed to contain the oldest track of a dinosaur thus far found and is dated at approximately 250 million years ago. Unless even earlier evidence is found, that means that any geologic formation that is older than 250 million years will not contain any dinosaur tracks because no dinosaurs existed before that time.

The three geologic deposits of the Glen Canyon group were formed between approximately 202 -170 million years ago. This time frame coincided with the rise of the early dinosaurs. These among others were the track-makers whose traces can be found at the Lake.

Next, you need a surface on that formation upon which the track makers footsteps and other traces will be preserved. The climate and conditions of the terrain during this time period were such that areas existed where tracks made in soft sand or mud could quickly dry and fill in with sand and other material preserving the original shape and structure of the imprint.

Once formed, the track site must have been deposited in a geologic environment that over time will preserve the site. The Glen Canyon recreation area is located within what is known as the Colorado Plateau, which from a geologic standpoint is a relatively stable area that has been spared much of the violent forces of nature which have deformed much of the land mass surrounding it.

Finally, the formation upon which the tracks were originally formed must again be exposed on the surface.

All of these factors have come together at Lake Powell.

Reading the Rocks

The terrain in and around the Glen Canyon National Recreation Area showcases more than 8,000 feet of a geologic record that represents about 300 million years of the earth's history.

Over time the area around Lake Powell has been a sea-floor, seashore, coastal plain, swamp, floodplain, and home to the largest most long–lived sand dune desert in the history of the world. Tracks and other traces of creatures have been found in rock formations dating from the Permian time. (290 – 245 million years ago) to the Cretaceous (145 – 65 million years ago). This time sequence begins well before the rise of dinosaurs, spans through their evolution and ends at the time of their extinction.

This book focuses on tracks and traces made during a portion of the period near the end of the Triassic and encompasses the Lower Jurassic Period (209 - 170 million years ago). Geologically, this time is represented by three sedimentary rock formations collectively known as the Glen Canyon group. The individual formations are known as the Wingate Sandstone, Kayenta Formation and Navajo Sandstone.

The Glen Canyon Group

Wingate Sandstone

The oldest and bottom-most formation in the Glen Canyon Group, the Wingate Sandstone, is an ancient sand dune deposit varying in thickness from 100 - 400 feet. The beginnings of the deposition of the Wingate formation occurred at a time of great change in this part of the world. Prior to this time all of the world's land masses had combined together into the super-continent called Pangaea. This was surrounded by a single global ocean called Panthalassa.

For about 20 million years before the Wingate was deposited, this area consisted of big, slow moving river and stream systems, lakes and swamps. Plants and trees grew along the river banks and shorelines. Some of these trees would later be preserved and become the petrified wood which can be found at the Lake in what is known as the Chinle formation which lies directly below the Wingate. The dominant creatures during this time were large amphibians, relatives of crocodiles, other reptiles, and reptile-like mammals. Dinosaurs at this point were only slowly beginning to evolve and ranged in size from smaller chicken and turkey sized species, to the dimensions of large size dogs.

There came a time between 200 and 209 million years ago that both the climate and life changed dramatically. In the area of what is now the East Coast, the super-continent began to rift and split apart. This rifting was accompanied by massive volcanic eruptions that released huge amounts of lava and spewed vast amounts of sulfur dioxide and carbon dioxide into the atmosphere. The Palisades along the Hudson River in New Jersey and New York are some of the observable present day remnants of that event. It is estimated that this volcanic activity may have persisted as long as 500,000 years. Eventually, this rifting split off what would eventually become the European and African continents and formed the Atlantic ocean.

The mass extinction of over 75% of all life on earth occurred in approximately this same time frame. It is believed that this extinction was at least in part caused by the massive volcanic activity, resulting climate change and acidifying of the oceans.

The lighter coloration at the lower part of the Wingate signifies the high water mark of the Lake.

This climate change turned the once wet tropical area around Lake Powell into an arid, dry, windblown desert-like environment. There are areas just down lake from the Rincon that one can literally put their finger on this stark transition zone between the Chinle and Wingate formations.

The Wingate normally is exposed as steep cliffs. Tracks and other traces are usually only found on blocks that have fallen off the cliffs and split open. The occurrence of tracks in this formation increases the higher up one goes. One explanation for this might be that after the extinction event, it took a long time for land-based surviving populations to recover in any great numbers.

Depending on water levels, Wingate Sandstone at Lake level, can be found around the Rincon, Good Hope Bay, and Cha Bay on the San Juan arm of Lake Powell.

Kayenta Formation

Another dramatic change in the environment marked the beginning of the accumulation of the next deposit, the Kayenta formation. The climate again became wetter in the region and supported rivers, streams, and shallow lakes. Over time this accumulated sand deposits to a depth of between 250 to 330 feet deep

Many tracks sites are located near the transition between this layer and the Navajo sandstone formation which lies directly above it. One possible reason for this might be that at the time of the transition, the environment was again returning to a windblown arid state. This might have made for better conditions for the preservation of tracks.

Depending on water levels, this deposit can be seen at Lake level south of the Escalante arm, Good Hope Bay, and the Great Bend area of the San Juan arm on Lake Powell.

Wingate
Transition
Chinle

Navajo
Kayenta
Wingate

Navajo Sandstone

At the time the Navajo sandstone was being deposited, the climate again became hot, windy and arid. What is present-day Utah was then located about 15° north of the equator. Wind blown sand accumulated to depths of 1,170 - 1,230 feet and covered an area over 5000 square miles.

Recent research suggests that the origin of much of the sand may have been from the erosion of the Appalachian Mountains. Until recently, this deposit was thought to be a lifeless sand dune desert. Discoveries of petrified wood, tracks, and remnants of playa lakes and springs revealed that a number of different types of ecosystems existed within this desert environment.

Periods of rain, possibly seasonal monsoons, and periodic high levels of groundwater, are now believed to have created conditions conducive to the formation of a food chain sustaining a variety of life during this period. Interestingly enough, the higher one goes in this formation, the larger the dinosaur tracks one finds. This may correspond with the fact that as dinosaurs evolved over time they grew in size.

Navajo sandstone covers approximately 1/3 of the surface area of the Glen Canyon National Recreation Area. Tapestry Wall south to the Rainbow Bridge and the Escalante Arm all contain classic exposures of this formation.

Navajo
Sandstone

What made the tracks ?

Determining the exact species of a specific track maker is not something that can be done with any degree of certainty. It has been estimated that less than 10% of all dinosaur types that have ever existed had been identified by paleontologists, which leaves more than 90% that have left no fossil record of their existence

Many fossilized tracks have been found that cannot be linked to any known dinosaur. Additionally, the feet of many track makers contained fleshy structures that are not apparent from merely observing their preserved foot bones. For these reasons, paleontologists do not attempt to classify tracks by the species of animal that made them. Rather, they classified tracks as to track types.

Typically, track type names are derived from some physical characteristic of the track or by the location of its first discovery. The most common track types found in the Glen Canyon group are : *Anomoepus, Grallator, Eubrontes, Otozoum, Batrachopus, and Brasilichnium.*

Anomoepus

Anomoepus is a track type believed to have been made by Ornithopods. These were a group of bird–hipped dinosaurs that started out as small, bipedal herbivores. They evolved in size and were the forerunners of the duck billed dinosaurs. The front feet are smaller and the hind feet have three toes. The toes of *Anomepus* are shorter and wider compared to the long, narrow shape of Grallator. The early Ornithopod dinosaurs were only about 3 feet long.

Anomoepus *Discovered by Steve Dzelzgalvis*

Grallator

Grallator represents a small slender-toed footprint ranging in size from as little as an inch to over 6 inches in length. Often times the track will display rounded pads and delineating joints between the bones of the toe. The track maker is believed to be a small dinosaur that walked by means of its two rear limbs.

A possible candidate for being a track maker is a theropod called *Coelophysis*, whose fossilized skeletons have been discovered in great numbers at a quarry on Ghost Ranch near Abiquiu, New Mexico. This meat eating dinosaur had a long slender neck and tail. Adults ranged in size from 7 to more than 9 feet in length and have a height of up to 3 feet at the hip.

Coelophysis *Exhibit - Ruth Hall Museum of Paleontology at Ghost Ranch*

Grallator

Eubrontes

Eubrontes represents a large tract type thought to be made by carnivorous theropod dinosaurs. The track width is greater than *Grallator,* and narrower and longer than *Anomoepus.* Track sizes vary generally up to 18 inches, although a cast measuring almost 21 inches has recently been discovered up-lake from Bullfrog marina on Lake Powell.

One possible *Eubrontes* track maker is a carnivorous theropod called *Dilophosaurus*. Adults reached 18 feet in length, weighed 600 to 700 pounds and held their head 6 to 7 feet off the ground. Three fossilized remains of this dinosaur were discovered in the Kayenta formation near Tuba City, Arizona.

Exhibit - The Dinosaur Museum Blanding, UT

Dilophosaurus

Otozoum

Otozoum tracks are thought to be made by a large biped, possibly a prosauropod. These were a group of early herbivorous dinosaurs that had a long neck and small head, fore limbs shorter than hind limbs, and a large thumb claw. The tracks are around 8 inches wide and maybe more than 12 inches long.

Otozoum

Batrachopus

Batrachopus is a track type believed to have been made by a small early Jurassic relative of the crocodile that adapted to living in more arid conditions. The tracks are generally small, typically several inches in length.

Batrachopus

Brasilichnium

Brasilichnium is a small 4 to 5 toed track ranging in size from less than an inch to 3 inches, with a wide foot and short toes. It is believed to have been made by mammal–like reptiles, possibly Tritylodonts or other animals similar to them.

Evidence suggests that they were herbivorous, lived in burrows, and were possibly nocturnal. Some scientists believe that mammals rose from this group. These were a very successful group even though they lived right beneath the ruling dinosaur's feet.

Brasilichnium

Crouching Dinosaurs

Traces of dinosaurs in a crouched or squatting position are rare. Apparently, when some theropod dinosaurs stopped walking, they sat back on their ankles or crouched, much like a modern day ostrich. Whether this behavior was for the purpose of resting or some other reason is unclear.

Before I met Dr. Lockley, I had never heard of crouching traces. In 2005, I showed him the block where I discovered my very first track at the Lake. I thought it was special because it appeared as if a dinosaur had stopped walking for some reason and it's tracks were preserved in a standing position. Martin studied that slab for quite some time but said nothing. Then he proceeded to tell me that yes, this was a very special find.

There was a crouching trace on the slab at about 10:00 o'clock up from the left big track near the top edge of the slab. Although I had visited and photographed that slab countless times before I had not recognized it for what it was. Had it not been for that experience, I never would have recognized the subsequent croucher traces later found at the Lake

When I asked him about what I thought were the standing dinosaur tracks, he smiled and said that the big track on the right was from an animal walking in one direction, it's preceding step was off the slab, as was it's next step. The left track, possibly made by the same animal at a different time, or another animal altogether, is traveling in a different direction and it's preceding and next steps are also both off of the slab.

This first crouching trace was discovered in the spring of 2005. At that time only about five were known in all of North America and this was one of only two found in the western United States. This also was the only one found in the Kayenta formation. In all, this slab contained: the crouching trace of a *Grallator* track type, walking tracks of a *Eubrontes* track type, and the swimming tracks of a possible small crocodile. The original slab is now again below Lake level as a result of rising water.

Detail of Croucher trace from upper left portion of slab on preceding page

Before the slab was again submerged by the rising water, a mold was taken to document the tracks. A latex mold consisting of several layers of latex, with a layer of burlap sandwiched between the layers was prepared.

After drying, the mold was peeled off the slab. The resulting mold can then be filled with a special plaster to form an exact replica of the surface of the slab.

I discovered a possible second crouching trace in the fall of 2006. This slab contained the trace of a crouching Eubrontes. While not yet verified by paleontologists, preliminary review of the track and photographs suggest that the crouching trace may well be the largest of its kind in the world.

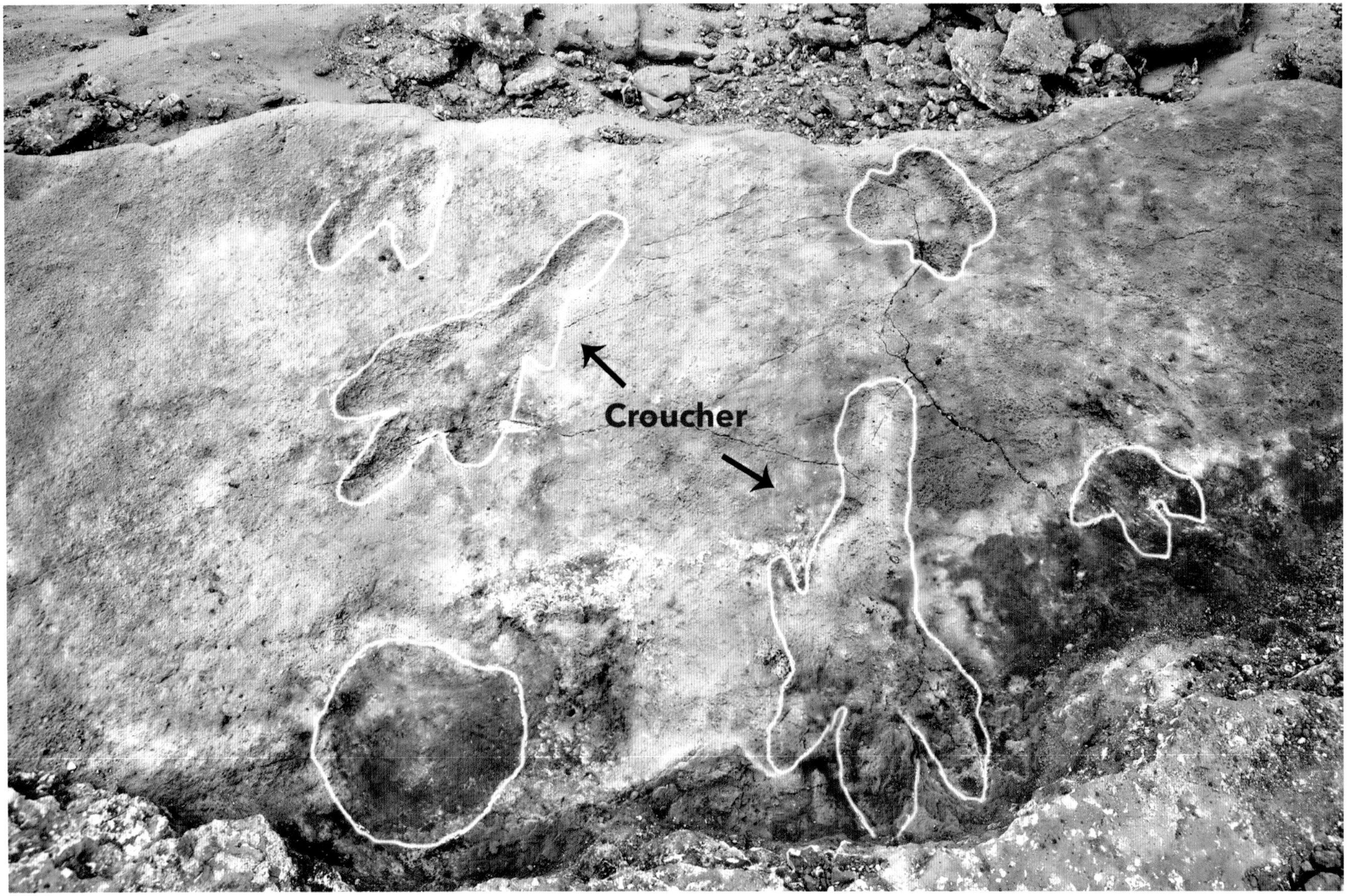

Traces outlined for emphasis

A possible third croucher trace was found in the spring of 2008.

Images of this block have been forwarded to paleontologists who confirm it is a possible croucher, however, formal study and positive confirmation have yet to be made.

Since the 2005 croucher discovery, other traces have been found in China, Poland and one near the Arizona - Utah border.

If Paleontologists agree these other two are indeed crouchers, then that means that Glen Canyon National Recreation Area has the greatest concentration of theropod croucher tracks anywhere in the world with a total of three.

Rising Lake levels at the first croucher find

Tracking Dinosaurs

Next to having a sharp eye and a cursory understanding of geology, an important attribute to successful tracking is keeping an open mind. A common mistake that novice trackers make is to have a preconceived visual image of what they are looking for.

Typically, this would be searching for a large three toed impression in the ground. While this is the classic representation of a dinosaur track in the minds of the majority of the general public, in the real world that image only represents a small fraction of the type of tracks and traces that dinosaurs have left behind. Tracks come in a variety of shapes, sizes, and are found in differing states of erosion.

Another common perception is that dinosaurs and their tracks were all huge. Actually, it is estimated that over a fourth of all known dinosaurs weighed less than 220 pounds, about the same as a large adult human. The gigantic dinosaurs represent only about 15% of the known animals. Dinosaur tracks in the Lake Powell area range in size from less than 1 inch to over 20 inches.

Tracks present in various states of preservation depending upon the substrate in which they were formed and the effects of erosion, among other things.

Tracks can be found as impressions

As freestanding casts

As a raised track (cast) on rock

In flattened relief

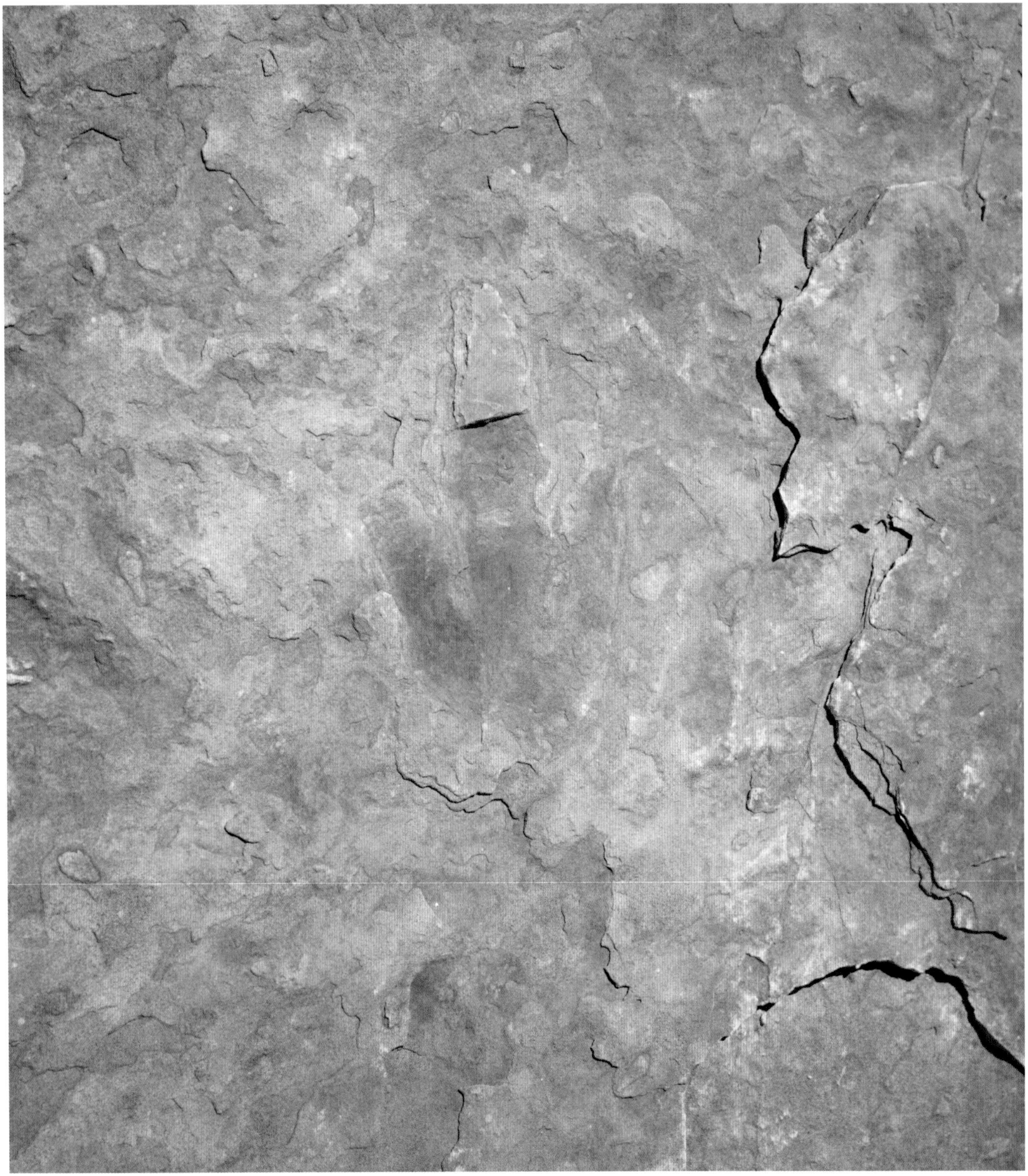

To just a mere
shadow on a rock.

Ghost tracks or under-tracks are created by the compaction of the material underneath an actual track. After that track has eroded away, the subsurface under the track provides residual clues to its former presents.

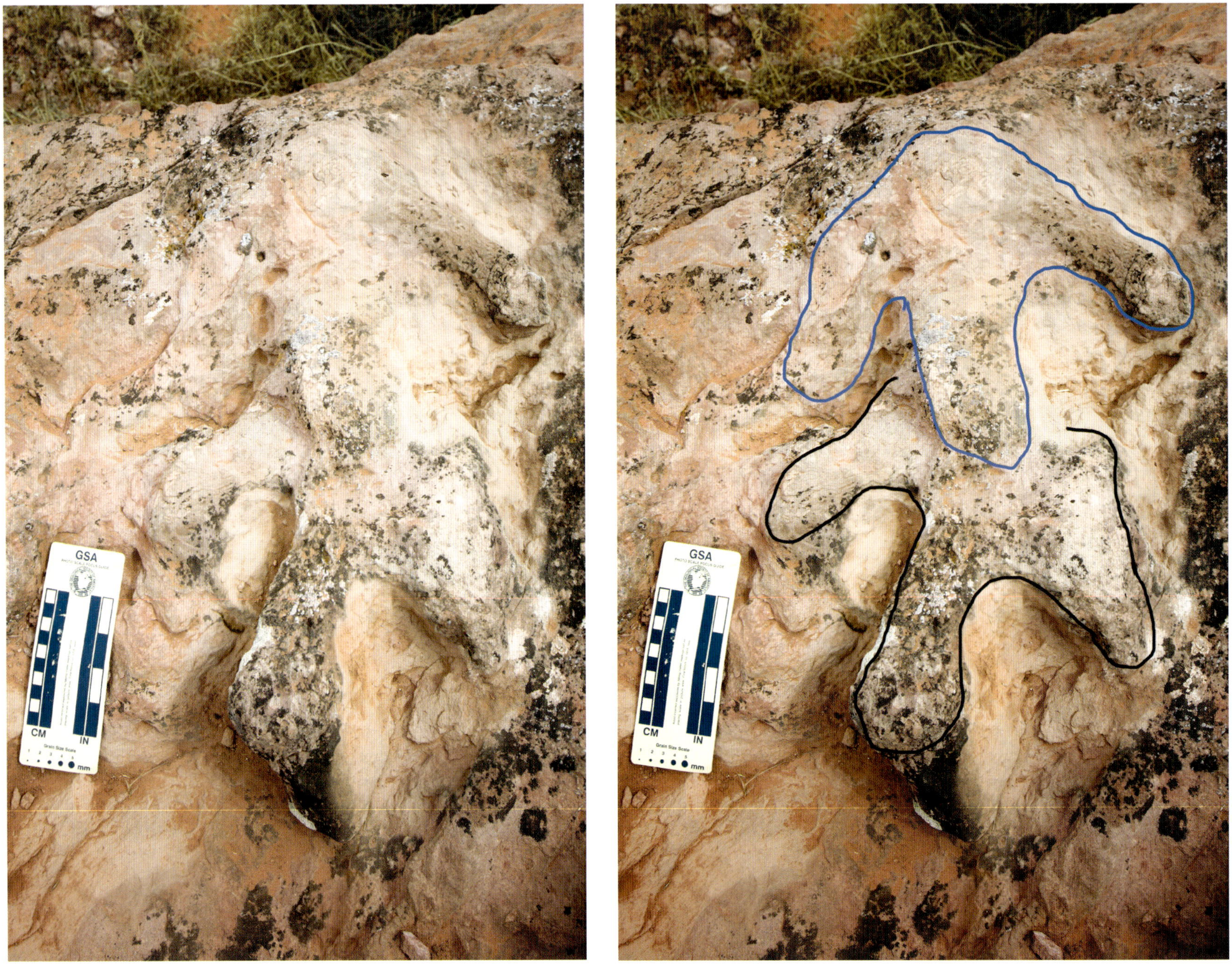

Sometimes tracks are found overlapping each other, this is called over-stepping.

Grallator over-stepping *Batrachopus* at upper left

Tracks can also present in a variety of colors

And be found in various states of erosion

Distortion

Environmental conditions at the time the track was made, intervening geologic processes and erosion may become factors in distorting a track's appearance. Sometimes, to the point where one wonders if what they are looking at is really a track.

GSA
CM
IN
mm

So, how does one go about determining if what is being observed is really a track or not? It's not an exact science, but some common sense may point you in the right direction.

Assuming you are in looking in a geologic formation that is age and type appropriate to contain dinosaur tracks, my personal method is to first assign an arbitrary status to a find. I use a three tier process.

If there is no question that what I am looking at is a track, I call that a **definite**.

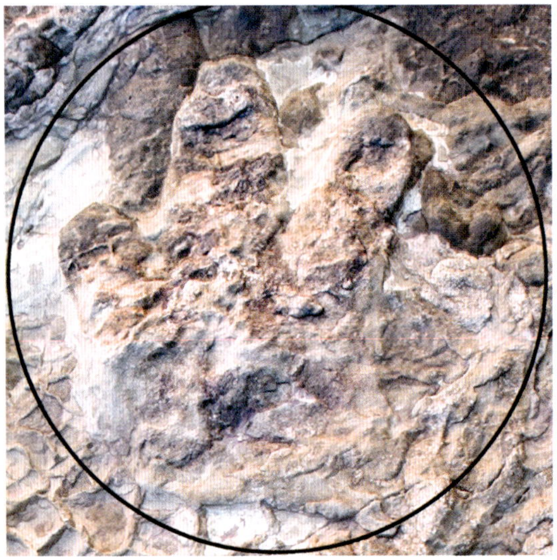

If it has some characteristics of a track, such as appropriate size and general shape and other clues, I call that a **probable**.

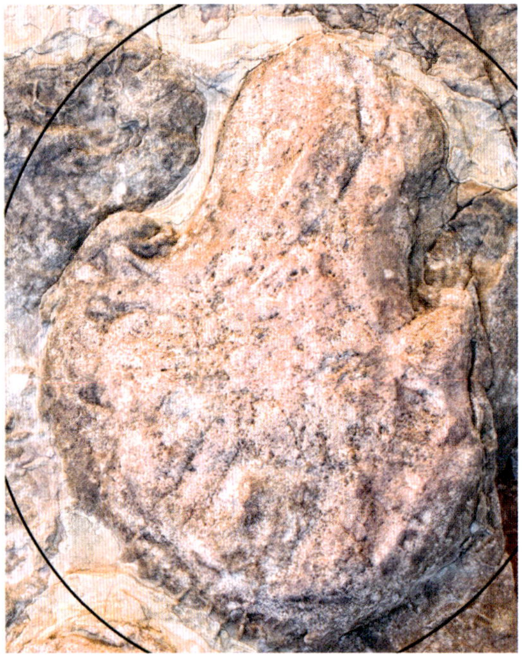

If it is a real toss up and I can't definitely say one way or the other, I call it a **possible**.

I will elevate the status of **possible** or a **probable** to the next higher level if I locate a **definite** in the same area. A **definite** will tell me that a track-maker was present, and that the ground it was walking on was conducive to track preservation.

If no **definite** is found, I tend to place much less weight on a **possible** as being an actual track. I'll also look at a **probable** with a little more skepticism.

Here, # 1 is a **definite**.

Alone, # 2 would be a **probable**; with the presence of #1, I'd elevate it to also being a **definite**.

Alone, I'd call # 3 a **possible**; being in proximity to #2, I'd elevate it a **probable**.

With the presence of #1 on the same slab, I'd make # 3 a **definite**.

1 and # 3 seem to have been made by the same animal. The tracks are the same size and shape and they are both oriented in the same general direction.

Additionally, the distance between the two tracks seems appropriate for a single stride given the size of the track.

Tracks are not only found by looking down, sometimes they will be above you.
This well known trackway is located beneath an overhang high up on a sheer Navajo Sandstone wall.

Navajo Sandstone wall

Depending upon the water level, there are other sites on the Lake one can drive a boat under and look up to observe tracks.

Tracks can be found in the most unexpected places.

GSA
PHOTO SCALE/FOCUS GUIDE
CM
IN
Grain Size Scale
1 2 3 4 5 mm

Several sites at Lake Powell contain a concentrated mass of tracks which I call a trample site. These sites may represent the last wet places around a drying waterhole. Track makers possibly would concentrate in this area to seek the benefits of a vanishing water source.

Often, there are so many overlapping tracks that it is difficult to distinguish individual tracks or even recognize them as tracks.

Most every irregularity on the massive block above is a *Grallator* track or track remnant.

This image was recently taken in Audubon Park in New Orleans. Many bird species over-winter there. Notice the similarity between this large concentration of bird tracks around a drying puddle and the Grallator trample site.

This trample site contains a mixture of Eubrontes tracks.

Batrachopus Tracks

Documentation

When I first started finding dinosaur tracks, the thrill for me was in the discovery. After I found one, I was mainly just interested in finding the next one. Searching for tracks with Dr. Lockley introduced a whole new perspective. There was much to be learned from each individual find. The actual discovery of a fossilized track or trackway is really only the beginning of a much larger process.

To obtain the most scientific value from a discovery, a process of documentation is undertaken. Much like modern day crime scene investigation, thoroughly investigating and documenting the track site leads to a better understanding of the nature of the track maker. Knowing what geologic strata the track was formed in provides a way to infer the age of the track. Taking measurements of the track's dimensions provides clues as to the track maker's size. Tracks suggest how many functional toes the animal had as well as other physical characteristics.

Trackways

Paleontologists refer to a number of tracks made by the same animal in sequence as a trackway. Trackways provide even more information about the track maker. Measuring the distance between strides it is possible to estimate the animals height at the hip, it's normal travel gait, and preference for walking, running or hopping. Measuring the angle between the steps and the width of the trackway provide insight into the track makers body proportions. Clues as to its preference for habitat, possible diet, and social behavior can also be extrapolated from the trackway data and its surrounding matrix.

Eubrontes trackway

Possible *Brasilichnium* trackway

Without documentation, much of the scientific value of a track or track site can be lost. If an undocumented track site is disturbed or removed from its original surroundings, accurate dating becomes difficult or impossible, and other valuable information about the track maker can no longer be determined

The documentation process involves several steps. One of the first is establishing the location of the find. Taking and recording a GPS (global positioning system) reading fixes the location, permitting it to be found again. One difficulty in using hand-held GPS units at Lake Powell is that it requires a line of sight to at least three different satellites to get a precise position. If the track site is located in a steep walled canyon or too near a canyon wall, only a general position can be obtained. In such cases, a journal entry giving a general description of the location may be all that is possible. In addition to location information, a description of the rock strata bearing the track should also be recorded

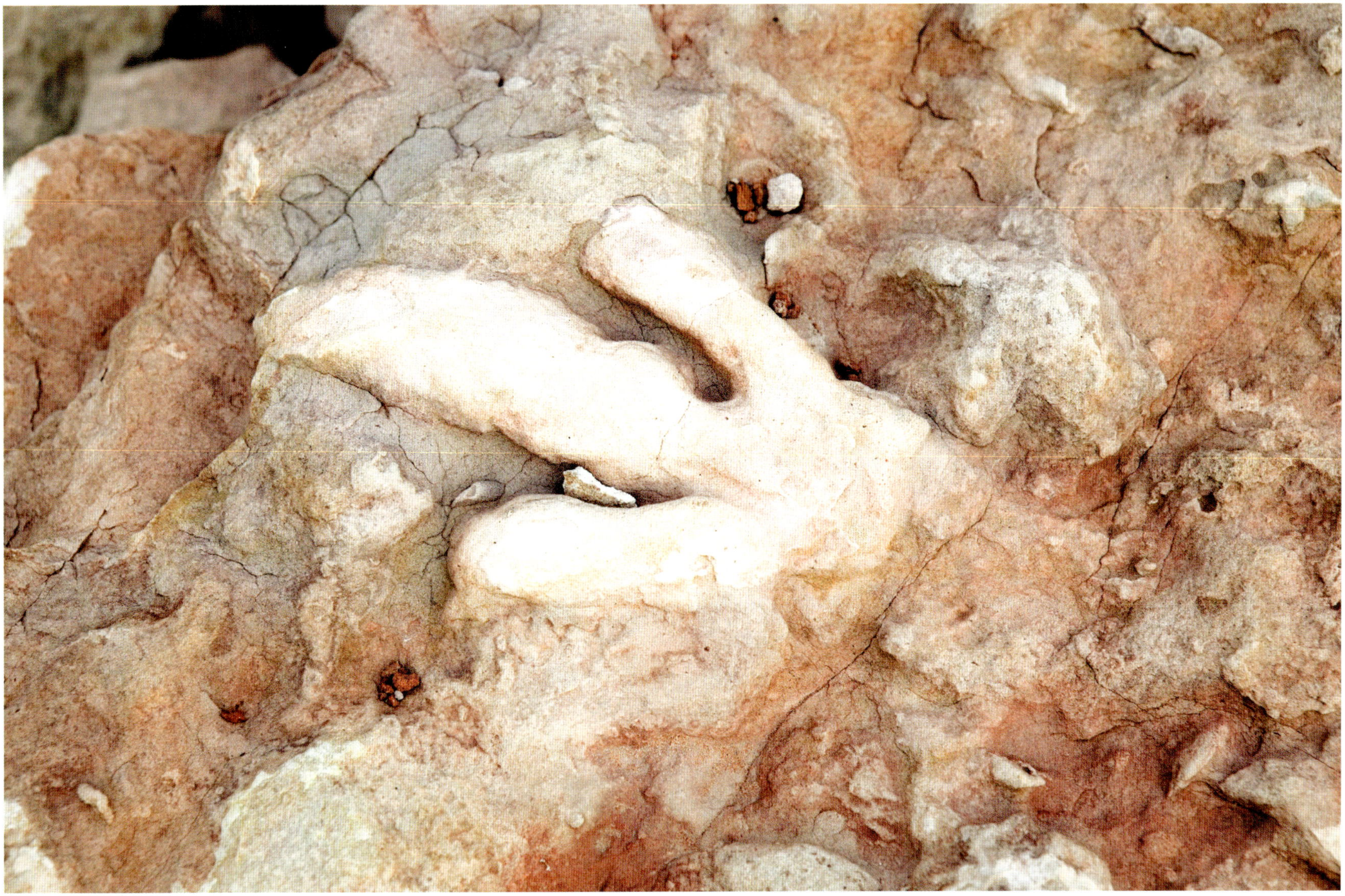

Photographing

Photographing tracks can often present a challenge due to less than ideal lighting conditions in the field and the often obscure nature of the tracks themselves. Frequently, when you photograph a track and later review the image, the results are less ideal than anticipated. The best lighting for photographing tracks is in the early morning and late afternoon or evening. The sun at these times is at a lower angle in the sky and creates shadows and contrast which makes the tracks more apparent. In the real world though, field work is done throughout the day and in all kinds of weather.

Paleontologists have used a number of methods to highlight tracks photographed in marginal lighting conditions. If the track is an impression, water can be poured into it to provide contrast. While this brings the viewer's eye to the tracks, it can also distort their true shape.

This is especially true for tracks found at Lake Powell where most are found in sandstone. The nature of sandstone is such that water tends to migrate and to be absorbed leaving a less than accurate representation of the true dimensions of the track.

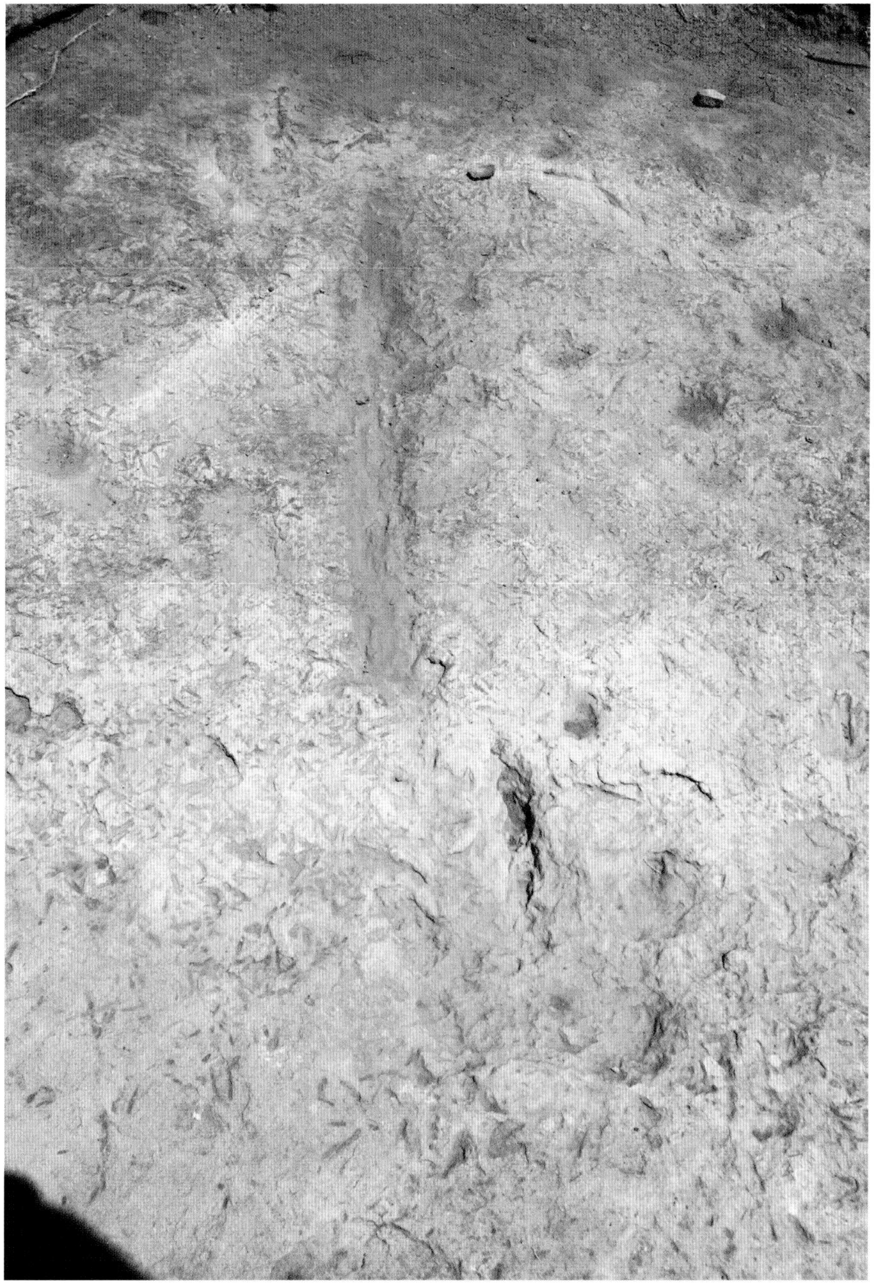 

Another method is to fill the track impression with sand or gravel of a different color than the surrounding rock to highlight its shape. Again, while visually interesting, this method covers the actual track and in reality we're only seeing a vaguely accurate outline of the track. If the track contains obscure or potentially important features such as skin impressions or detailed outlines of the digits, these will be covered up by the over lying sand.

The use of chalk to outline tracks is another way of highlighting them. Some paleontologists have an almost uncanny ability to flesh out extreme detail in tracks they encounter.

With the coming of age of digital photography and post processing, a new variety of methods are available to enhance track images. Now, it is possible to make the tracks more apparent, while accurately preserving their detail.

One method is to desaturate the color of the matrix surrounding the track.

Another technique is to increase the contrast and detail of the track to bring it out from the background.

For detailed scientific study of tracks, 3 dimensional imaging technology is now being utilized with amazing results.

To aid in making tracks more apparent to the untrained eye many of the images in this book have undergone enhancement for easier recognition.

When photographing tracks and traces it is also helpful to photograph the surrounding area. This helps in again finding the location in the event that GPS coordinates are inaccurate. This practice can also make it easier to identify the rock layer containing the track site.

Photographing track sites gives us the ability to review and study them more closely. It often happens that in the excitement of the moment one may just not notice or assign importance to what is right in front of them.

I had visited and photographed this site many times. It was only when preparing this image for use as a poster, that I noticed the set of smaller tracks just to the right of the middle toe.

This track was spotted while I was sitting on a nearby rock, having a lunch break.

The existence of the second track only became apparent when the image was later processed and more carefully examined.

This image was taken when I first began tracking. At the time I thought it was some sort of a plant trace. It was only after learning more about different track types and reviewing the photograph that I realized these were tracks.

Sometimes it happens that you start to explore an area that you think should hold tracks, but nothing is apparent.

After a long day of searching, with minimal results, I came upon this slab. The track in the upper third portion, detailed above, was what first caught my attention.

Closer examination revealed some curious traces of track-type *Brasilichnium* (1). This mammal-like reptile was possibly running, as evidenced by the length of the stride and that one foot print appears to be overstepping the other in this two-stride sequence.

Still closer examination of the bottom of the slab revealed a possible cause of it's haste.

Was this *Eubrontes* track maker chasing his supper?

That scenario came into question with an even closer look at the slab. A 3rd track set labeled # 2 in the *Brasilichnium* trackway sits next to the *Eubrontes* track. It is not clear whether the *Eubrontes* overstepped the *Brasilichnium* track. If it did, then a pursuit situation is possible. If it doesn't, the track was likely placed at a different point in time, and there was no encounter.

We also have a potential new player in the scenario. The track labeled # 3 is on a direct collision course with the *Brasilichnium's*. In the end, perhaps it was this player that ended up with a full belly.

What first looked to be a one track slab, turned out to be a very interesting study.

Scale

Rulers, coins, lens caps, and other reference objects of a known size can be included in images to provide scale. The key is *known* size. Glove sizes vary from extra large to extra small, lens caps vary with the diameter of the lens. If you include a reference that can be a variable, it is best to measure and note the size of the reference material when an accurate measuring device becomes available.

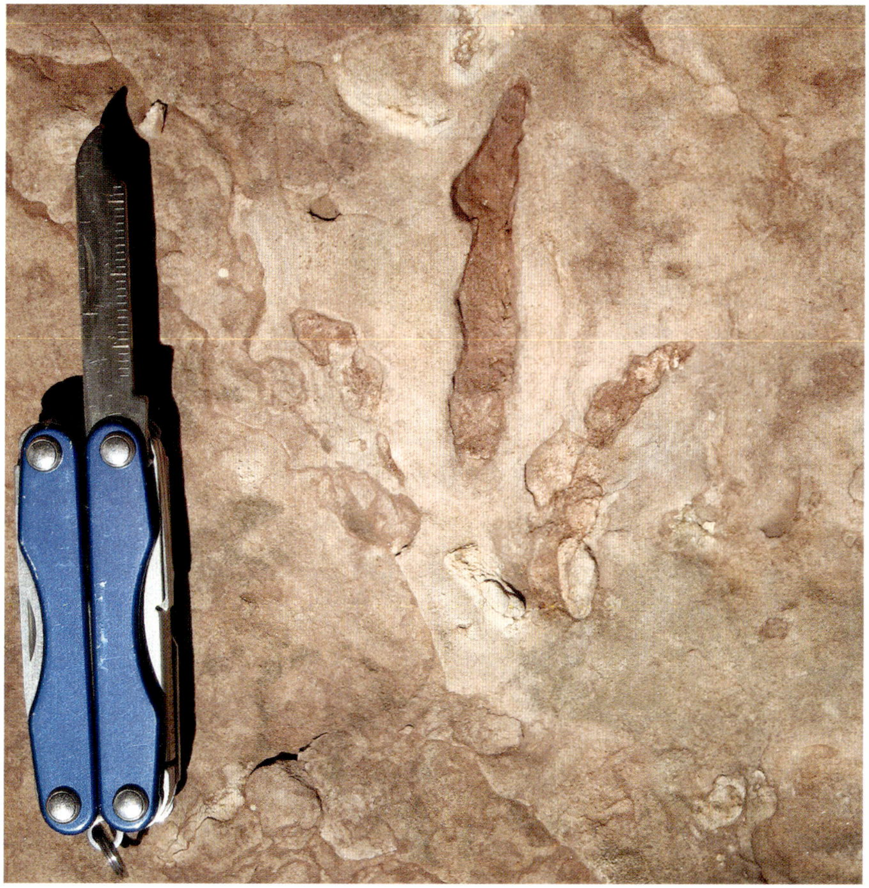

Mapping

When multiple tracks or a trackway is found, a mapping of the site may be undertaken. This involves laying out a grid with measuring tapes and accurately recording the placement of the tracks within the grid. The resulting data is then mapped onto graph paper to preserve an accurate record of the relationship and orientation of the tracks within the site.

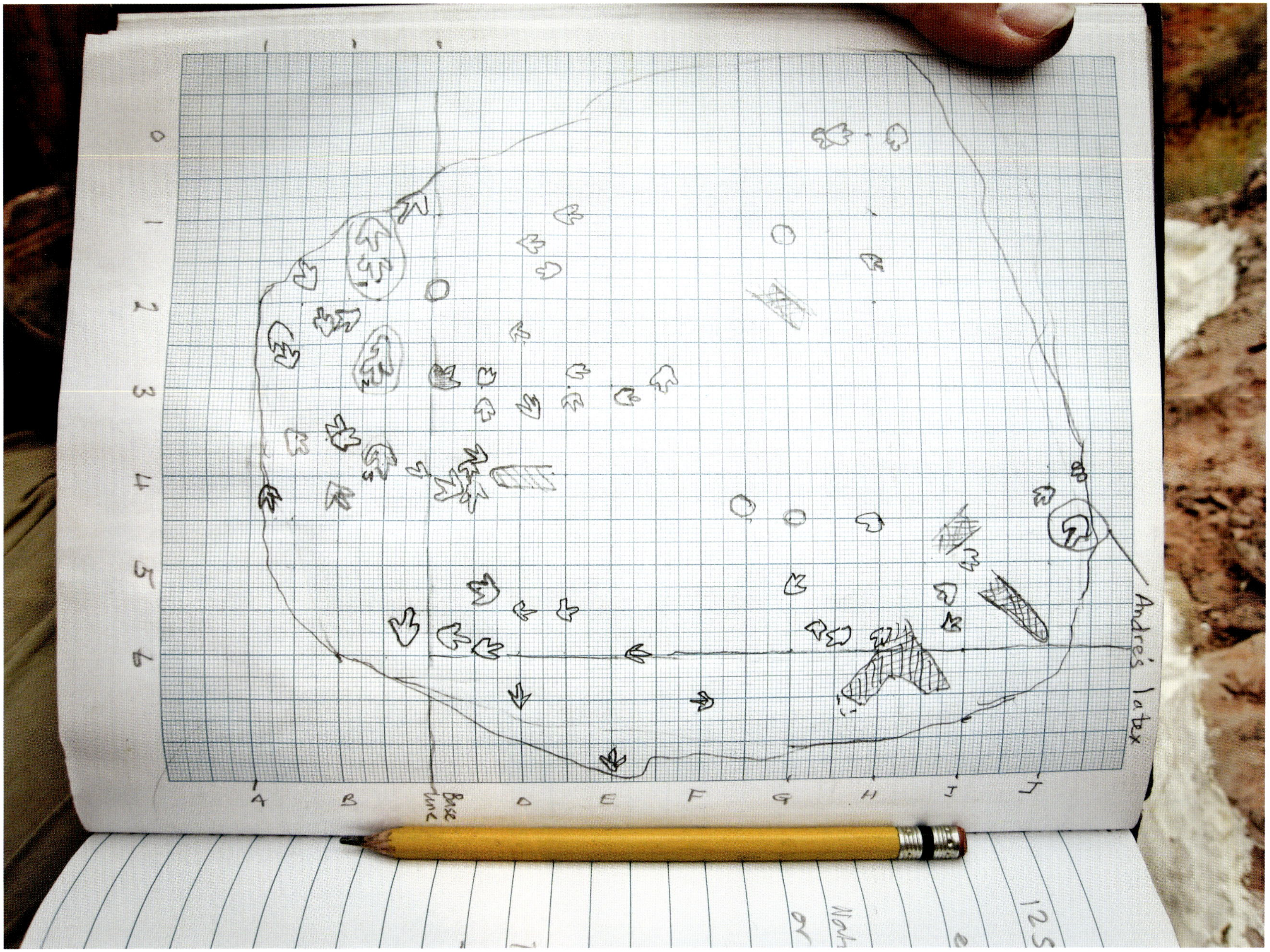

Mapped graph of a single slab containing more than 60 tracks.

Tracing

Tracing of the tracks or a trackway is another effective method of accurately recording the size, shape, and spatial relationship between the tracks. This involves placing a clear sheet of acetate over the tracks and tracing their outline with a Sharpie marker.

One interesting aspect of using acetate is that you can stick it on a white wall or other background just by creating a static charge in it. Once charged it will cling to the wall, is re-positionable and will adhere anywhere from days to weeks before requiring a recharge.

Latex molding

Exact replicas of tracks or a trackway may be made by preparing a latex mold. A special liquid latex is first brushed onto the track.

Several coats are applied and allowed to dry between applications. Burlap is then used to cover the latex to give it body.

When completely dry, the mold can be peeled off of the track without damaging it.

A special type of plaster is then poured into the latex mold. After curing, the latex is peeled off of the now hardened plaster and an exact replica of the original track is obtained.

In the past, people have poured plaster directly onto tracks thinking that this was the proper way to make a track replica. Not only does this practice often damage, deface or completely destroy tracks, it is also in **Violation of Federal Laws** enacted to protect and preserve these irreplaceable paleontology resources.

Special permits are now required to make latex molds of tracks on Federally protected lands, as even this process can damage tracks if done incorrectly.

The latex molding illustrated in the previous pages was done by experienced researchers under permit from the National Park Service.

At least one track site in the Glen Canyon Recreation Area has already been permanently damaged by individuals attempting to replicate a track-site using plaster. Apparently other track sites outside the GCNRA have also sustained damage due to this practice.

Journal entry - March 23, 2005
Ship's Log of the *Daina Marie*
by Martin Lockley

A dinosaur was walking
with his friend one day
and as they both were talking
one was heard to say……

"After fourscore years and ten
when we're both on our backs
will we be remembered
when there's nothing left but tracks "?

"Who knows what the future holds"'
the other said at last…..
"Perhaps they'll throw away the molds
from which you and I were cast"

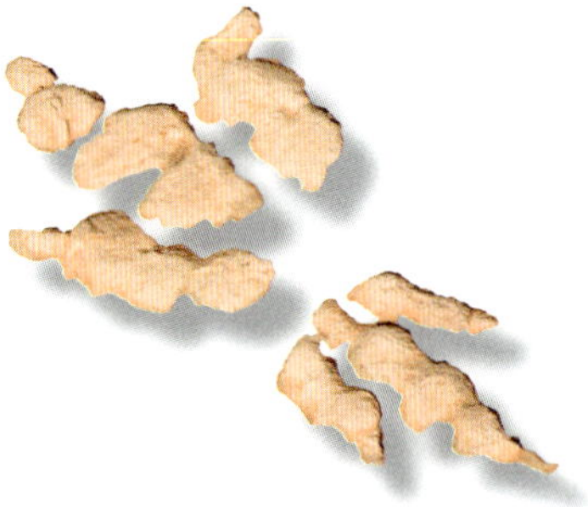

The Art of Seeing

My mother was a dentist. One of my favorite childhood memories was going to her office waiting room to read copies of Highlights Magazine. That was the one for kids with the page that contained a number of hidden images within a picture. Scrutinizing track site photos provides a somewhat similar experience.

To sharpen and train your tracking eye, the following pages contain unenhanced images of track sites with clues to position and orientation on the page following. You will probably spot some I may have missed.

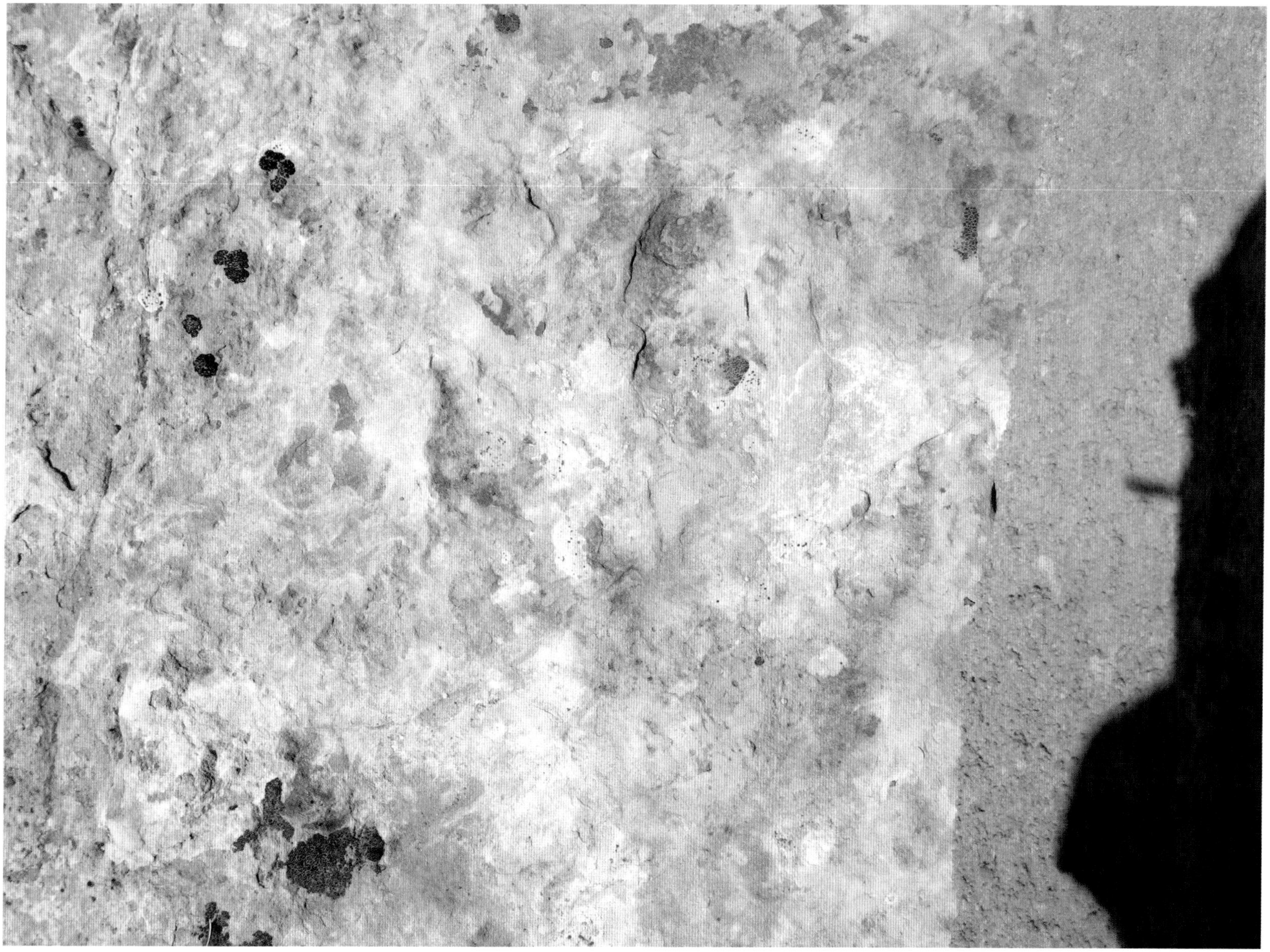

Find the track(s)

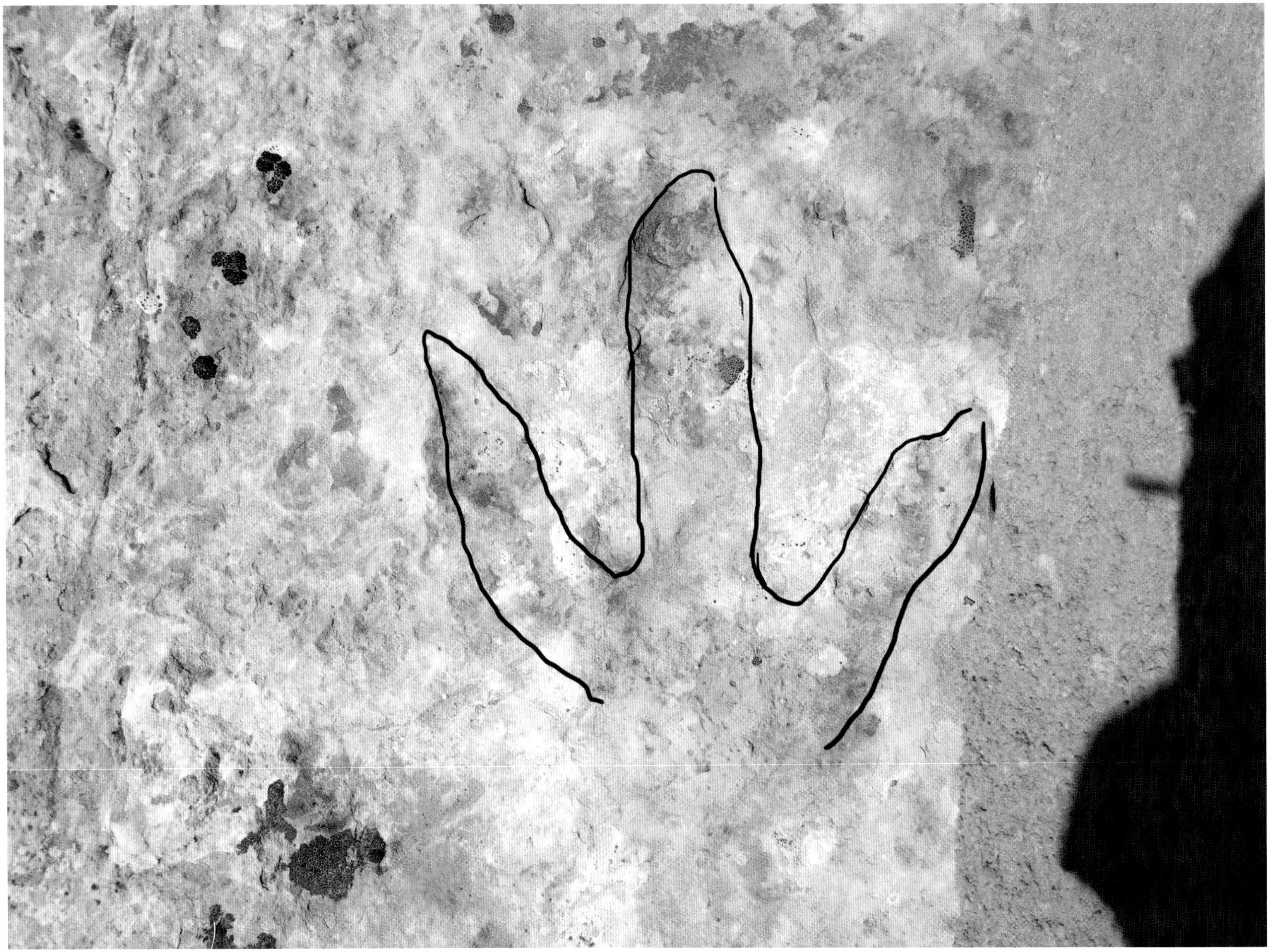

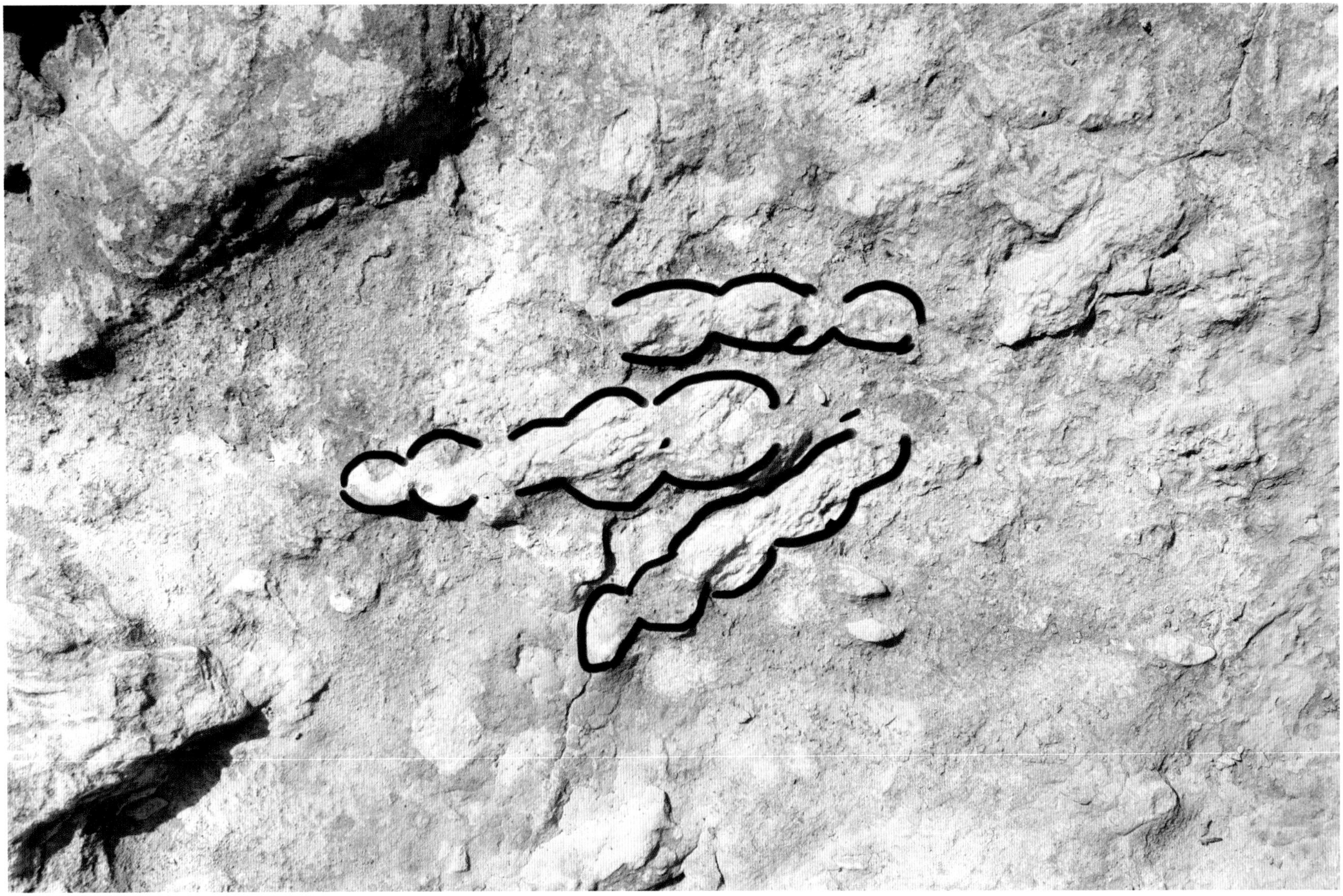

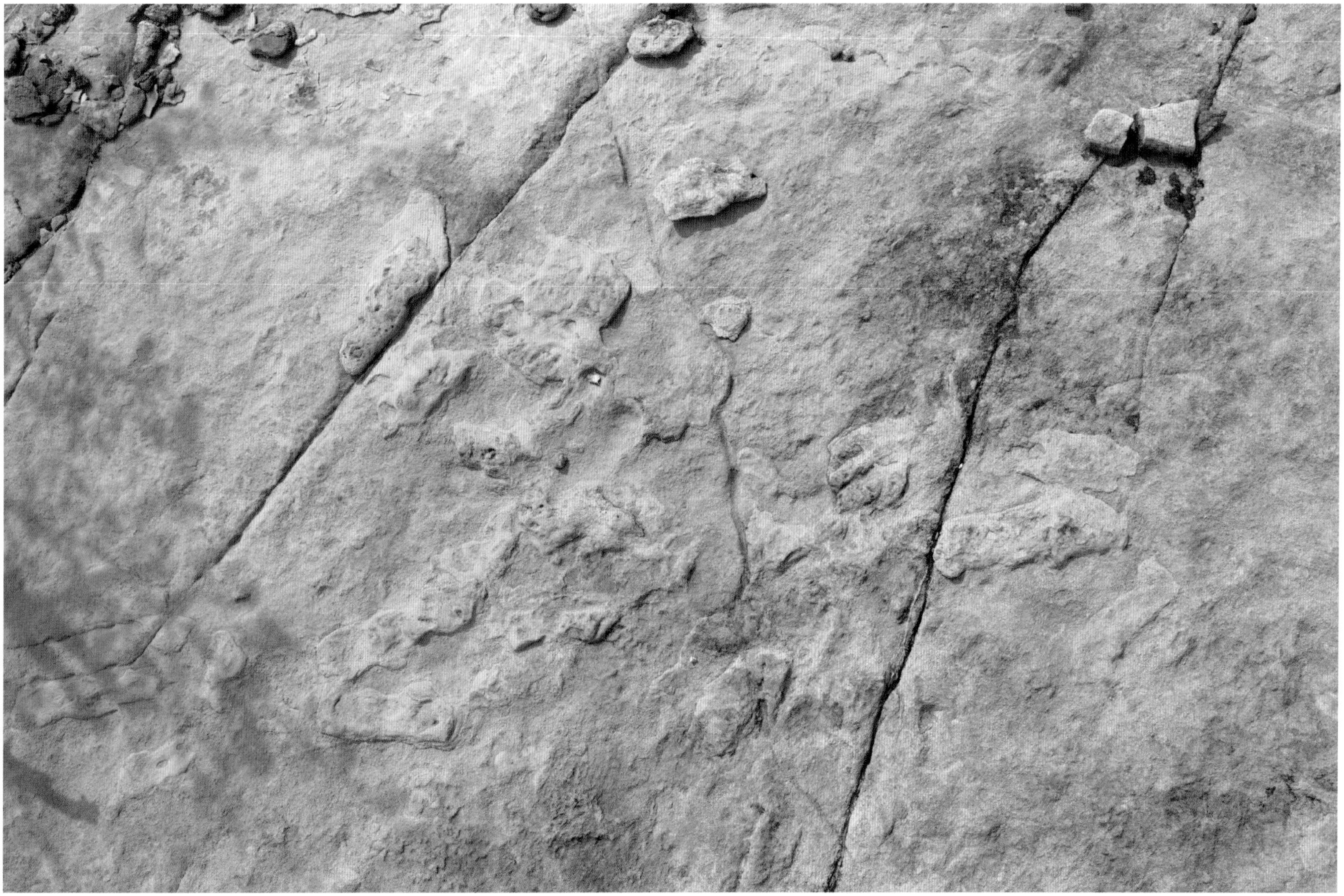

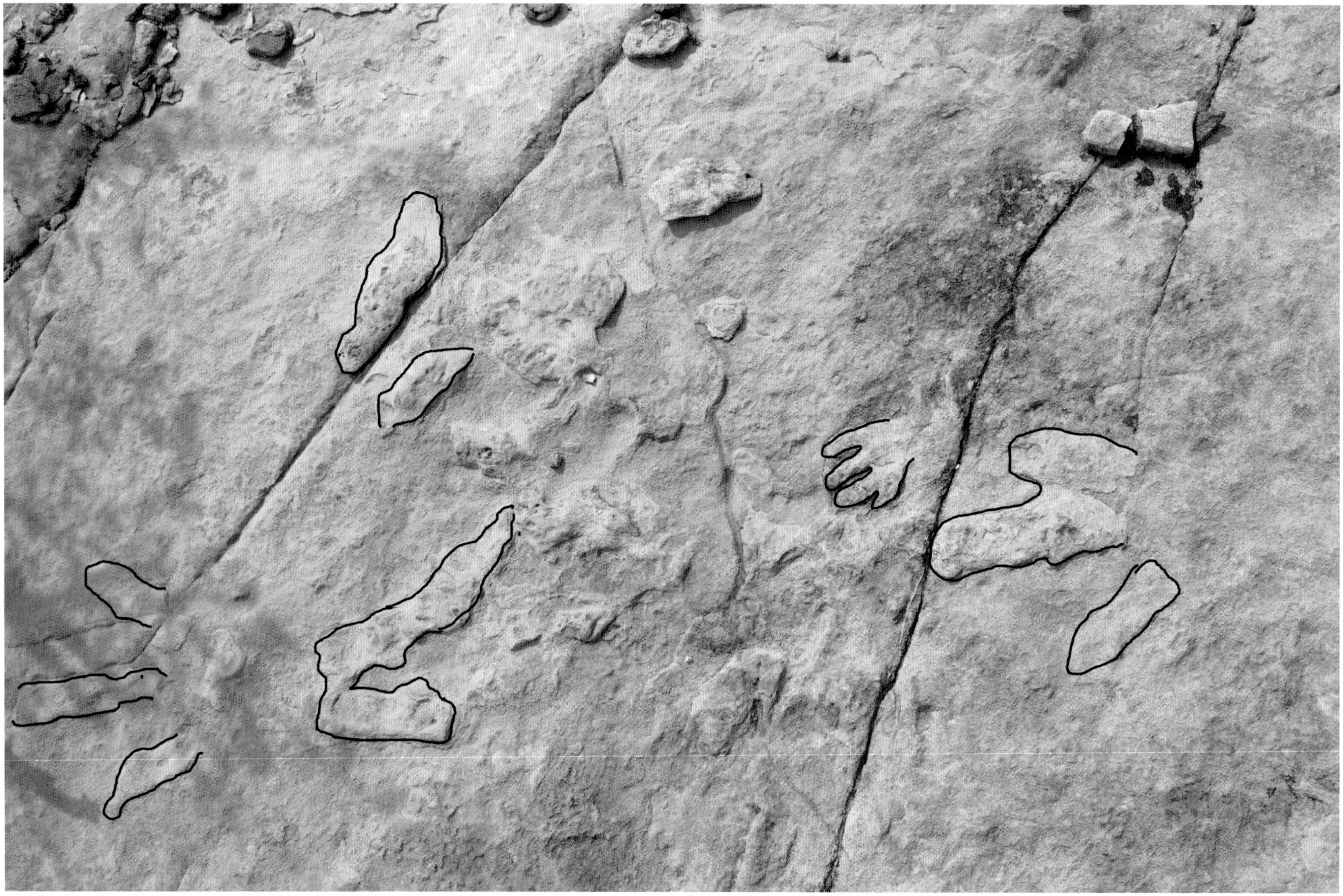

Other Traces

In addition to dinosaur tracks, other traces from the Jurassic can be found in the Glen Canyon Group.

During the times that river systems existed in the area, large colonies of clams thrived on the river bottoms. Occasionally, great floods would scour the clam beds and pile them downstream.

Eventually they would solidify into masses called Coquinas. This block is probably in the area of 190 - 200 million years old.

Lake Powell is now beginning to be plagued by a new invasive species called quagga mussels. In reality the region has a history of clams and related species dating back nearly 200 million years.

Finding petrified wood in the Navajo Sandstone formation at Lake Powell is not all that common. Normally, you would find it in the Chinle formation which underlines the Wingate Sandstone

This is one of the few sites that I've found containing both petrified wood and dinosaur tracks together in Navajo sandstone.

The site seems to be the top surface of a playa. I believe this represents the end-stages of an ancient oasis as it is filling in and drying up. Trees and other plant life were beginning to grow on its drying surface, and dinosaurs were roaming it's shores.

The majority of track sites at Lake Powell contain tracks of predatory meat eating carnivores. What was the food source that allowed this population to survive?

Recent research suggests that some dinosaurs were swimmers. Fish in the streams and playas may have been a food source. Evidence of this swimming behavior can be found as traces of single or parallel grooves where the swimmer's feet scratched the bottom.

Some of the smaller vertebrates dug burrows which they used for shelter, raising young and sometimes for hibernation. Mammal-like reptiles probably stayed in the borrows during the day for protection from predators and relief from the heat.

On occasion these borrows would fill-in and be preserved. Examples of these can be found in the Glen Canyon group.

Possible insect pupae in Navajo Sandstone

While similar in appearance to a borrow tunnel complex, paleontologists have interpreted this as a root system remnant.

This root system may have been infested with Jurassic age relatives of termites.

Mud cracks and other natural erosional processes create these interesting patterns

Possible coprolite (Dino poop) although unusual to find in the Glen Canyon Group. This specimen is firmly attached to the sandstone matrix below it.

Tracking dinosaurs at Lake Powell is not without its hazards. If you are camping with a houseboat or on shore in a tent, choose your spot wisely. Flash floods during the monsoon season, typically between July to September, can create heavy downpours resulting in spectacular waterfalls. These are wondrous to look at but not much fun to be camped under.

Rock slides are another situation to be aware of. Camping in areas where the Chinle formation meets the Wingate Sandstone can be especially dangerous. As the weaker Chinle erodes, support for the Wingate above disappears and a rock-slide may occur.

The potential for heat stroke and dehydration, are other factors that need to be addressed. If tent or boat camping with a generator, be aware of the risk of carbon monoxide poisoning. Scrambling around boulders is best done with sturdy boots to protect ankles from sprains.

I always wear gloves to protect my hands from abrasion and a possible encounter with thorns or scorpions. You might also come across a tarantula, but for the most part they look more dangerous than they really are.

Snake encounters at the Lake, in my experience, are not all that common. If you do happen upon one of the diminutive rattlers sometimes found on the rock pile securing your anchor line, they usually alert you to their presence before you notice them. Most times, they will want to avoid you as much as you want to avoid them. The Bull snake pictured above is a constrictor and not poisonous. It is a good source of rodent control and should be left alone.

You may be lucky enough to encounter an elusive ring-tail cat. Although not dangerous, they may come on board at night and steal any food scraps you may have left behind.

Acknowledgments

Most all of the discoveries illustrated in this book would not have been possible but for our houseboat the *Diana Marie* and our scout boat. Boats do not run by themselves. Over the last 12 years of tracking at the Lake I have been privileged to have had three of the best first mates any captain could hope to encounter.

Mac McCarty "The Professor" was my initial first mate on the *Daina Marie* and in addition to his dutch oven cooking skills, knot tying, and legendary fishing acumen is a first class Dino tracker in his own right.

Steve Dzelzgalvis, my nephew and godson was my second first mate. His insights into life, cooking skills, and guitar playing made every trip an event. He also has a number of Dino track discoveries under his belt.

Susan Delgalvis, the boat princess, is my lifetime first mate, cook extraordinaire, opera-on-the-Lake advocate, and an aggressive gaming partner as well as a no-nonsense tracker.

George Callison, retired paleontologist, and a very talented landscape artist, who provided me with the initial paleontology insights, contacts, and continued support has been a vital resource. Thanks also for proofing the first draft.

Dr. Martin Lockley, now recognized as one of the world's foremost experts in dinosaur tracks (and budding Poet), has been instrumental in giving me a deeper understanding of track identification and the vital role of the documentation process.

James Kirkland, State Paleontologist at the Utah Geological Survey, for his review and feedback on paleo burrows and other traces.

Andrew Milner, Paleontologist and Curator, St. George Dinosaur Discovery Site at Johnson Farm, for initial feedback on the possible croucher discoveries.

Alex Downs, Ghost Ranch Resident Paleontologist, for background on *Coelophysis,* and other early dinosaurs.

Boating on a body of water as large as Lake Powell does not come without its hazards. Much of my tracking time is spent in the off-season and in some of the more remote areas of the lake. When emergencies happen, situations can go from bad to very very bad in relatively short order. Many of the personnel at Aramark have gone out of their way to keep the *Daina Marie* and this project on an even keel.

First and foremost I'd like to thank Capt. Tony Anderson, a true gentleman's gentleman, of Aramark Executive Services at Bullfrog Marina for getting both me and the boats out of bad situations on more than one occasion. The same goes for Captain Whitey Klindt and the rest of the crew at the Executive Services dock.

Thanks also to the boat mechanics at the Bullfrog Boat Shop who keep the *Daina Marie* running. That crew has always been there for us when we needed them.

Also, thanks to the folks at Bullfrog Dry Storage for their excellent service.

Finally, I'd like to thank Flip Todd and Vered Mares of Todd Communications for guiding me through the publishing process.

The tale of this journey of discovery began with finally locating a legendary giant bullfrog ensconced on the side of a mountain. Perhaps it is fitting to conclude this portion of the adventure with a natural formation embedded in the side of a petrified Navajo Sandstone dune, which to some may appear to be a giant *Eubrontes* track. Maybe this section of Iceberg Canyon should become known as "Tracker's Cove".

Additional Reading:

A Guide to the Fossil Footprints of the World *by Martin Lockley*

Dinosaurs of Utah *by Frank DeCourten*

Ancient landscapes of the Colorado Plateau *by Ron Blakely and Wayne Rainey*

Jurassic West *by John Foster*

Dinosaur tracks and other fossil footprints of the Western United States *by Martin Lockley and Adrian Hunt*

Canyon Country Dinosaur Tracks and Trackers *by F. A. Barnes*

Footprints on the Shores of Time *by F. A.Barnes*

In the Page Arizona area, dinosaur tracks are on display at:

The John Wesley Powell Museum

Big Water Visitor's Center

Carl Hayden Visitor Center at Glen Canyon Dam